A Kid's Guide to Drawing™

How to Draw Cartoon Rodents

Curt Visca and Kelley Visca

The Rosen Publishing Group's
PowerKids Press™
New York

Dedicated to Kelley's mom, Alrene Allan, in honor of her favorite M-O-U-S-E

Published in 2003 by The Rosen Publishing Group, Inc.
29 East 21st Street, New York, NY 10010

First Edition

Editor: Natashya Wilson
Book Design: Kim Sonsky
Layout Design: Emily Muschinske

Illustration Credits: All illustrations © Curt Visca.
Photo Credits: Cover photo (porcupine) p. 20 © Animals Animals/Joe McDonald; cover photo and title page (hand) © Arlan Dean; pp. 6, 10 © Animals Animals/Robert Maier; p. 8 © Animals Animals/Breck P. Kent; p. 12 © Animals Animals/J. & P. Wegner; p. 14 © Animals Animals/Darek Karp; p.16 © Animals Animals/Ken Cole; p. 18 © Animals Animals/Patti Murray.

Visca, Curt.
How to draw cartoon rodents / Curt Visca and Kelley Visca.— 1st ed.
Editor: Natashya Wilson Book Design.
p. cm. — (A kid's guide to drawing)
Includes bibliographical references and index.
Summary: Provides information about various types of rodents and step-by-step instructions for drawing them.
ISBN 0-8239-6161-3
1. Rodents—Caricatures and cartoons—Juvenile literature. 2. Cartooning—Technique—Juvenile literature. [1. Mammals in art. 2. Cartooning—Technique. 3. Drawing—Technique.] I. Visca, Kelley. II. Title. III. Series.
NC1764.8.R64 V58 2003
741.5—dc21

2001004497

Manufactured in the United States of America

CONTENTS

Cartoon Rodents

Did you know that there are more rodents in the world than any other living **mammals**, including people? Rodents make up half of all mammals. There are more than 2,000 different **species** of rodents. They live all around the world in every type of **habitat**, including oceans, rivers, deserts, mountains, and the Arctic.

All rodents have a pair of front teeth that grow throughout their lives. They have to **gnaw**, or chew, on hard objects to wear down these teeth. The scientific name for rodents, Rodentia, comes from the Latin word *rodere*. It means "to gnaw."

Most rodents are fairly small. The largest rodent is the capybara, which lives in South America. It looks like a large, hairy pig. It can grow to 4 ½ feet (1 m) long and can weigh 110 pounds (50 kg). The smallest rodent is the pygmy jerboa, which lives in the deserts of Africa and Asia. It looks like a tiny kangaroo and is only 2 inches (5 cm) long.

In this book, you will learn about eight different types of rodents and how to draw a cartoon of each one. The Terms for Drawing Cartoons list on page 22 can help you learn the drawing shapes.

Remember that a cartoon isn't meant to look just like its real-life subject. When you draw a cartoon, you include only the basic lines and shapes. You can **exaggerate** parts of the rodents to make your cartoons funny. Let your creativity run wild!

You will need the following supplies to draw cartoon rodents:

- Paper
- A sharp pencil or a felt-tipped marker
- An eraser
- Colored pencils or crayons to add color

To draw cartoon rodents, find a desk, a table, or another quiet place with lots of light. Keep all of your supplies nearby. Take your time, and practice your cartoons. Soon you'll be a great cartoonist!

EEK! It's a mouse! Oh, it's only a cartoon. Turn the page to see what all the excitement is about.

Enjoy drawing cartoon rodents!

The Mouse

If you have ever found packages of food in your kitchen that have been gnawed on, you might have a mouse living in your house! Mice will eat any food they can find. They will even eat soap and glue. Mice live in fields, in people's houses, and in other buildings.

Mice can be from 3 to 14 inches (8–35.5 cm) long from nose to tail. They have gray, brown, or white fur. Mice use their sharp claws for **grooming**, scratching, and climbing. House mice like to build nests out of paper, clothing, and anything else they can chew up. Mice are called pests, because they can do a lot of damage to farmers' crops and to buildings. They can also spread **diseases**. However, many people enjoy keeping **tame** mice as pets.

1

Let's begin by drawing a circle for the first eye. Draw a letter *C* for the second eye. Place a dot in each eye.

2

Start by the eyes and make a long letter *U* with a short straight line at the end. Draw a slightly curved line for the jaw. Make an oval for the nose, and shade it in. Add whiskers.

3

Good work! Start above the eyes and draw a curved line for the top of the head. Make two letter *C* shapes for the ears. Add a curved line inside the ear on the right.

4

Draw the front legs with curved lines. Use wiggly lines for the toes. Make a straight line for the belly. Draw the back legs with curved lines and add wiggly lines for toes.

5

You're incredible! Start from behind the ears and draw a long curved line for the back. Make the tail using two wavy curved lines that connect in a point.

6

Add dots and action lines to the mouse. Draw lines for the ground and small ovals for food crumbs.

The Rat

The **population** of house rats in the United States is probably equal to the population of people! Rats first lived in Asia. From there they climbed aboard ships and traveled all around the world, getting off wherever the ships **docked**. Now rats live everywhere. Some species are kept as pets.

As mice do, rats have sharp, strong teeth. They can gnaw through wood and can even chew holes in lead pipes. Rats can get into storehouses and damage food, such as grain, because of their powerful bites. Rats can also carry diseases that make people and animals sick. Rats look like mice but are larger. They can be from 6 to 23 inches (15–58 cm) long, not including their long tails, which range from 3 to 8 inches (8–20 cm) long.

1

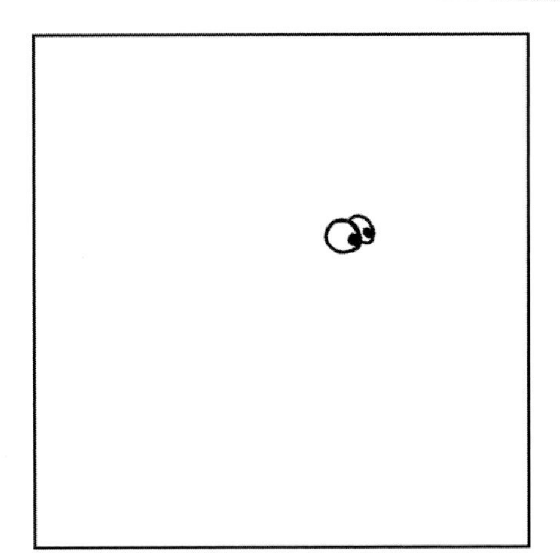

Start by drawing a circle for the first eye and a backward letter *C* for the second eye. Add a dot in each eye.

2

Make a letter *J* with a short line at the end for the head and mouth. Add a short curved line where the *J* curves. Shade in a circle for the nose. Draw the jaw, then make whiskers.

3

Amazing job! Draw a wiggly line for the top of the head. Make an upside-down letter *U* for each ear.

4

Draw curved lines and wiggly lines to make the legs. Add thin letter *U*'s for the toes. Use wiggly lines to draw the belly.

5

Fantastic work! Make a long, curved wiggly line for the back. Draw two long, wavy curved lines that connect for the tail.

6

Add lines and dots on your rat. Draw rectangles and dots to show the ground. Make a water bottle using curved lines and rectangles. Wow!

The Hamster

Hamsters are small, round rodents with black eyes and fur that ranges from white to brown. They have **pouches** inside their cheeks in which they store food. Hamsters eat grain, fruit, and vegetables. They are from 2 to 11 inches (5–28 cm) long and can weigh up to 32 ounces (90 g). Hamsters live in the wild in Asia and Europe, but they also have been **domesticated**. Many people keep hamsters as pets, because they're tame and easy to care for. Wild hamsters live in **burrows** that they dig underground. These burrows have several rooms that are connected by tunnels. Some rooms have nests for sleeping. One room is always used to store grains and food. In the winter, wild hamsters **hibernate**.

1

You'll begin by drawing a circle for the first eye. Make a letter *C* for the second eye and add a dot in each eye.

2

Start by the eyes and add a wiggly line. Shade in a small oval for the nose. Make the mouth with a letter *U* and a letter *C*. Draw a wiggly line for the jaw. Add whiskers.

3

Draw a curved wiggly line above the eyes for the head. Make a backwards letter *C* for the front ear and a short curved line for the back ear. Add a curved line inside the front ear.

4

Draw the legs with wiggly lines and curved wiggly lines. Make toes out of thin letter *U*'s. Add a wiggly line for the belly.

5

You made it look easy! Draw a long, curved wiggly line from the front ear to the back foot to finish the body of your hamster.

6

Add dots and small lines on your hamster. Draw rectangles and dots for the ground. Use curved lines, rectangles, and ovals to make a water bottle. Add action lines.

The Guinea Pig

The guinea pig gets its name because when it is really excited, it **squeals** like a pig! Guinea pigs are not really pigs. They are rodents that come from Guyana, South America.

A guinea pig's fur can be black, reddish, cream, and all shades of brown. Their fur can be long, short, straight, or curly, depending on the species.

Guinea pigs often are kept as pets. If you have a male and a female guinea pig in your house, expect lots of babies, called **pups**. Guinea pigs begin to **breed** when they are two months old. Females can give birth up to five times each year. They have an average of four pups in a litter. That means it's possible for a guinea pig to have 20 pups every year!

1

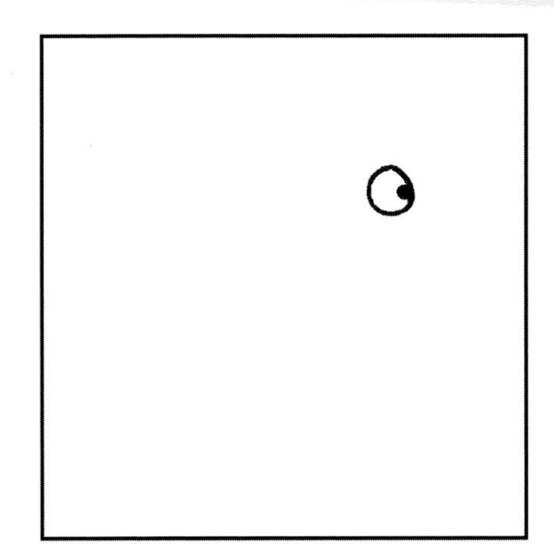

You'll begin by making a circle with a dot inside for the first eye.

2

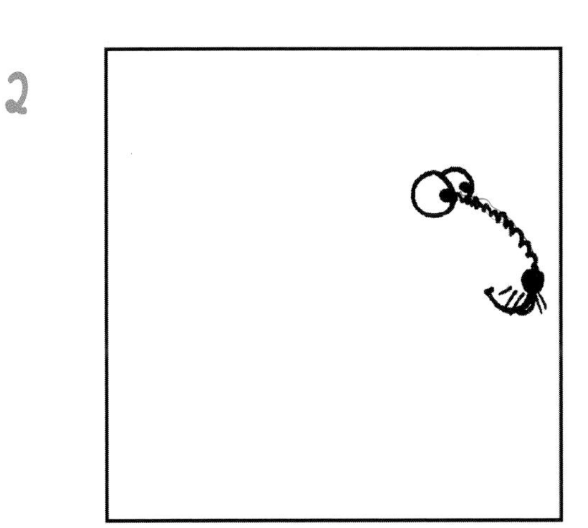

Make a curved line and a dot for the second eye. Add a wiggly line. Shade in an oval for the nose. Draw a letter *U* and a short line for the mouth. Add whiskers.

3

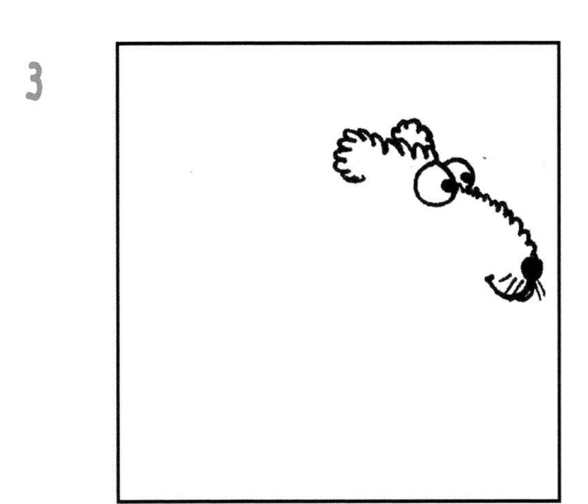

Splendid work! Start above the eyes and draw a curved wiggly line for the head and the front ear of your guinea pig. Make another curved wiggly line for the back ear.

4

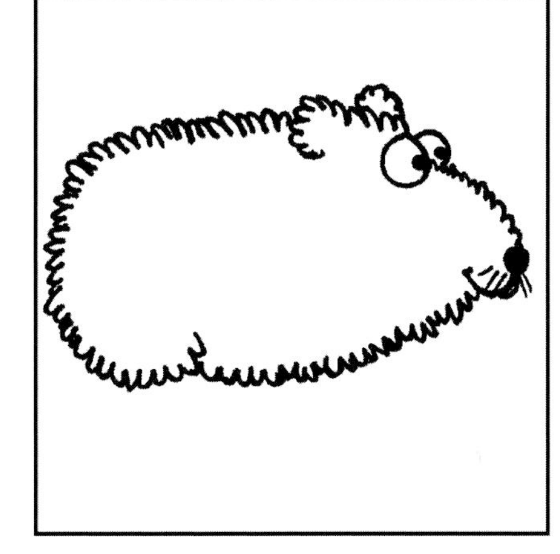

Draw a long, curved wiggly line for the back and the bottom of the guinea pig. Add another wiggly line for the belly and the neck.

5

Great work so far! Make the feet using curved lines and thin letter *U*'s.

6

Make dots and more wiggly lines on your guinea pig for detail. Draw some action lines. Add small rectangles and dots for ground cover. Super!

The Beaver

The phrase "busy as a beaver" might come from the fact that beavers are always working to build **dams** and **lodges**. They use their sharp front teeth to cut down trees. Beavers work together to dig

canals to float the logs to a stream. There they use logs, rocks, and mud to build a dam. A dam can measure 5 feet (1.5 m) high and 10 feet (3 m) long. The dam stops the water and makes a pond, where the beavers build their lodge. Lodges are made of sticks, grass, and mud. They have underwater entrances. Beavers are also known for their large, 1-foot-long (30.5-cm-long) tails. Their tails help them swim. Beavers will slap their tails on water to warn other beavers of danger.

1

Start by making two circles for the eyes. Add a dot in each eye. Shade in an oval for the nose. Draw two curved lines for each ear.

2

Draw two letter *U*'s to form the mouth. Add two rectangles for the beaver's big front teeth. Make the rest of the head using curved wiggly lines.

3

Your hard work is paying off! Make two curved wiggly lines on the left and two on the right for the front legs. Draw thin letter *U*'s at the end of the legs for the paws.

4

Draw a very short wiggly line below each front paw. Make each back foot using thin letter *U*'s and a curved line. Add a wiggly line between the feet for the belly.

5

You're on your way to becoming a rodent expert! Draw a long curved line for the beaver's tail.

6

Use different shapes to decorate your beaver. Make logs using curved and straight lines. Show water with wiggly lines. Use upside-down letter *V*'s to make mountains. Beautiful!

The Squirrel

The tree squirrel is one of more than 200 species of squirrels. Tree squirrels often are seen **scurrying** up and down trees in parks and forests. They use their sharp claws to hold on to the tree trunks. Their bushy tails help them to steer as they jump from branch to branch. Tree squirrels use their front paws like hands to eat foods such as fruits, nuts, worms, and insects.

The flying squirrel is another interesting species. Flying squirrels have flaps of skin that connect their front and back legs. When a flying squirrel leaps from branch to branch, it spreads out its legs and skin to glide through the air. Although they don't really fly, flying squirrels can glide 150 feet (46 m)!

1

You'll begin your squirrel by making a circle for the first eye. Add a dot in the eye.

2

Draw a letter *C* and a dot for the other eye. Add a curved line. Shade in an oval for the nose. Make the mouth using two curved lines. Add whiskers and a line for the jaw.

3

Awesome! Draw a curved line for the top of the head. Make the ears out of upside-down letter *U*'s. Add a short curved line inside the ear on the right.

4

Make the front legs using curved lines. Use wiggly lines for toes. Add a slightly curved line for the belly. Draw a letter *C* to start each back leg. Add wiggly lines for the toes.

5

Draw a curved line to make the back. Add a long, wavy curved line for the squirrel's tail.

6

Draw dots and small letter *U*'s on the squirrel. Add action lines. Make the ground using wiggly lines and dots. Draw acorns using ovals and zigzag lines. Add a tree limb with leaves!

The Chipmunk

Chipmunks have reddish-brown fur and white and dark brown stripes on their faces and backs. There are about 17 different species of chipmunks. They live in North America, Europe, and Asia. Chipmunks make a squeaky "chip" sound, which is how they got their name. They eat nuts, seeds, berries, and insects. They use their large cheek pouches to carry extra food to their burrows, to be stored for winter. Like wild hamsters, chipmunks dig their burrows underground. A burrow can be more than 11 feet (3.5 m) long. It usually has several entrances and extra rooms for storing food. Chipmunks stay mostly in their burrows during the winter but sometimes come out on warmer days.

1

Begin by drawing a circle for the first eye and a backward letter *C* for the second eye. Add a dot in each eye.

2

Start by the eyes and make a letter *J*. Add a short straight line at the end for the mouth. Draw a curved line for the jaw. Make an oval for the nose, and shade it in.

3

Draw a short curved line above the eyes. Add an upside-down letter *U* for each ear, with a smaller upside-down letter *U* inside the front ear. Make another curved line below this ear.

4

Draw slightly curved lines for the front legs. Make the paws using small letter *U*'s. Draw a curved line for the belly. Add the back legs by drawing more curved lines and letter *U*'s.

5

Outstanding! Start from the head and add a long curved line for the back. Draw a long, curved wiggly line to make the tail.

6

To complete your chipmunk, use dots to make stripes. Add action lines. Use curved and wiggly lines to draw rocks and trees. Wonderful work!

The Porcupine

Porcupines are known for their needle-sharp **quills**, which cover their backs, sides, and tails. Common porcupines have about 30,000 quills! When a **predator**, such as a coyote, threatens a porcupine, the porcupine makes its quills stand on end. The porcupine then uses its tail to hit the enemy. The tail's quills stick into the other animal. More quills may fly out if the porcupine shakes itself. Quills can blind or even kill the predator. They can be up to 1 foot (30.5 cm) long. They have **barbs**, or hooks, on their ends, which makes them hard to remove.

Porcupines live in North America, Europe, and Asia. They spend their days in burrows or hollow logs. At night porcupines climb trees to find food. They eat leaves, bark, fruit, and twigs.

1

Begin the last rodent in this book by making a circle for the first eye and a letter *C* for the second eye. Put a dot in each eye.

2

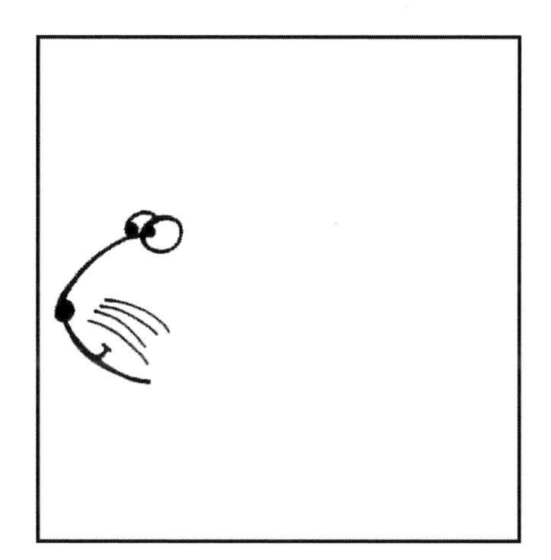

Draw a wide letter *C* for the front of the head. Shade in a small circle for the nose. Make a short curved line for the mouth. Add whiskers. I like your cartoon so far!

3

Take your time as you make numerous long, thin zigzag lines for quills on the back, the tail, and the side of your porcupine.

4

Fantastic job! Add more zigzag lines on the body to show more quills.

5

Make the front and the back feet using slightly curved zigzag lines. You're almost finished with your prickly porcupine!

6

Add more small quills. Draw wiggly lines and dots to show the ground. Make mountains using upside-down letter *V*'s, zigzag lines, and dots. Great work on your cartoon!

Terms for Drawing Cartoons

Here are some of the words and shapes that you need to know to draw cartoon rodents:

- Action lines
- Circle
- Curved line
- Dots
- Letter *C*
- Letter *J*
- Letter *U*
- Letter *V*
- Oval
- Rectangles
- Shade
- Straight line
- Wiggly line
- Zigzag lines

Glossary

barbs (BARBZ) Sharp spikes with hooks at the ends.
breed (BREED) When animals produce babies.
burrows (BUR-ohz) Holes dug in the ground.
canals (ka-NALZ) Waterways dug across land that are used to move objects or to carry water to places that need it.
dams (DAMZ) Strong barriers built across streams or rivers to hold back water.
diseases (duh-ZEEZ-ez) Illnesses or sicknesses.
docked (DOKT) To stop in a harbor to load or to unload cargo.
domesticated (duh-MES-tuh-kayt-id) Raised to live with humans.
exaggerate (ihg-ZA-juh-rayt) To make something bigger, better, or more important than it really is.
gnaw (NAW) To keep on biting something.
grooming (GROOM-ing) Cleaning the body and making it appear neat and tidy.
habitat (HA-bih-tat) The surroundings where an animal or a plant naturally lives.
hibernate (HY-bur-nayt) To spend the winter sleeping or resting.
lodges (LAH-jihz) Beavers' homes.
mammals (MA-mulz) Warm-blooded animals that have backbones and hair, and that give birth to live young. Dogs, horses, and humans are mammals.
population (pah-pyoo-LAY-shun) The number of animals that live in an area.
pouches (POWCH-ez) Skin that stretches to form baglike areas for carrying things.
predator (PREH-duh-ter) An animal that hunts other animals for food.
pups (PUPS) Baby guinea pigs.
quills (KWILZ) The long, sharp, thin spines on a porcupine.
scurrying (SKUR-ee-ing) Hurrying or running with quick, short steps.
species (SPEE-sheez) Animals grouped by things they have in common.
squeals (SKWEELZ) Makes a shrill, high sound or cry.
tame (TAYM) Gentle and unafraid of humans.

Index

Web Sites

To learn more about rodents, check out these Web sites:

http://falcon.jmu.edu/~ramseyil/kidspets.htm#G
http://griefnet.org/KIDSAID/petstuff/rodents.html